Around The World With

FOOD AND SPICES

Olives

Melinda Lilly

Rourke Publishing LLC
Vero Beach, Florida 32964

www.rourkepublishing.com

For my family, with love and thanks

PHOTO CREDITS:
Cover photo by Melinda Lilly • Artwork on title page: Detail of Corfu from Vonitsa. Edward Lear. Photo Courtesy of the Sotheby's Picture Library, London. • Page 7 Olive Orchard. Vincent Van Gogh. The Nelson-Atkins Museum of Art, Kansas City, Missouri (Purchase: Nelson Trust) 32-2 • Page 8 Illustration by Patti Rule • Page 11 Noah's Ark. Edward Hicks. Philadelphia Museum of Art: Bequest of Lisa Norris Elkins. 1950-92-7. Photo by Graydon Wood. • Page 12 Detail of Banquet Scene, Egyptian, c. 1400 B.C. The Nelson-Atkins Museum of Art, Kansas City, Missouri (Purchase: Nelson Trust) 64-3 • Page 15 Olive Picking. Attic Black Figure Shape: Neck Amphora. Painting by Antimenes Painter. From Vulci region: Etruria. 530 B.C.-510 B.C. © Copyright The British Museum. • Page 16 The Hope Athena, 2nd century A.D. copy after a Greek original of the late 5th century B.C.. School of Pheidias. Los Angeles County Museum of Art, William Randolph Hearst Collection. Photograph ©2000 Museum Associates/LACMA. 51.18.12 • Page 19 Kindling of the Hanukkah Lamp in a Polish-Jewish Home, by K. Felsenhardt. Poland 1893. Colored chalk and goache on paper. Kirchstein Collection HUCSM 66.79. From the HUC Skirball Cultural Center, Museum Collection, Los Angeles, CA. Photography by Lelo Carter. • Page 20 Detail of Jerusalem from the Mount of Olives. James Fairman. Photo courtesy of the Sotheby's Picture Library, London. • Page 23 The Prophet with His Companions in a Mosque: page from a manuscript of the Shahnama-i al-i Osman (Royal Book of the House of Osman) of 'Arifi, circa 1558. Turkey, Istanbul. Los Angeles County Museum of Art, the Nasli M. Heeramaneck Collection, Gift of Joan Palevsky. Photograph ©2000 Museum Associates/LACMA. M.73.5.446 • Page 24 The J. Paul Getty Museum, Malibu, California. Detail of Caeretan Hydria. Attributed to Eagle Painter. c. 525 B.C. Medium: terracotta. H. 44.6 cm; Diameter (rim): 22.9 cm.; Diameter (body): 33. • Page 27 Leonardo da Vinci, self-portrait • Page 28 Detail of San Gabriel Mission., c. 1835. Ferdinand Deppe. Oil on canvas, 27 x 37 inches. LAM/OCMA Art Collection Trust, Gift of Nancy Dustin Wall Moure. Photo by Chris Bliss. • Page 30 Photo by Jayne McKay

ILLUSTRATIONS:
Artwork on world map and additional illustrations by Patti Rule
Artwork on cover by James Spence

EDITORIAL SERVICES:
Pamela Schroeder

Library of Congress Cataloging-in-Publication Data

Lilly, Melinda
Olives / Melinda Lilly
p. cm. — (Around the world with food and spices)
Includes bibliographical references and index.
ISBN 1-58952-044-0
1. Cookery (Olives)—Juvenile literature. 2. Olive—Juvenile literature. [1. Olive.] I. Title.

TX813.o4 L55 2001
641.3'463—dc21

00-054317

Printed in the USA

Table of Contents

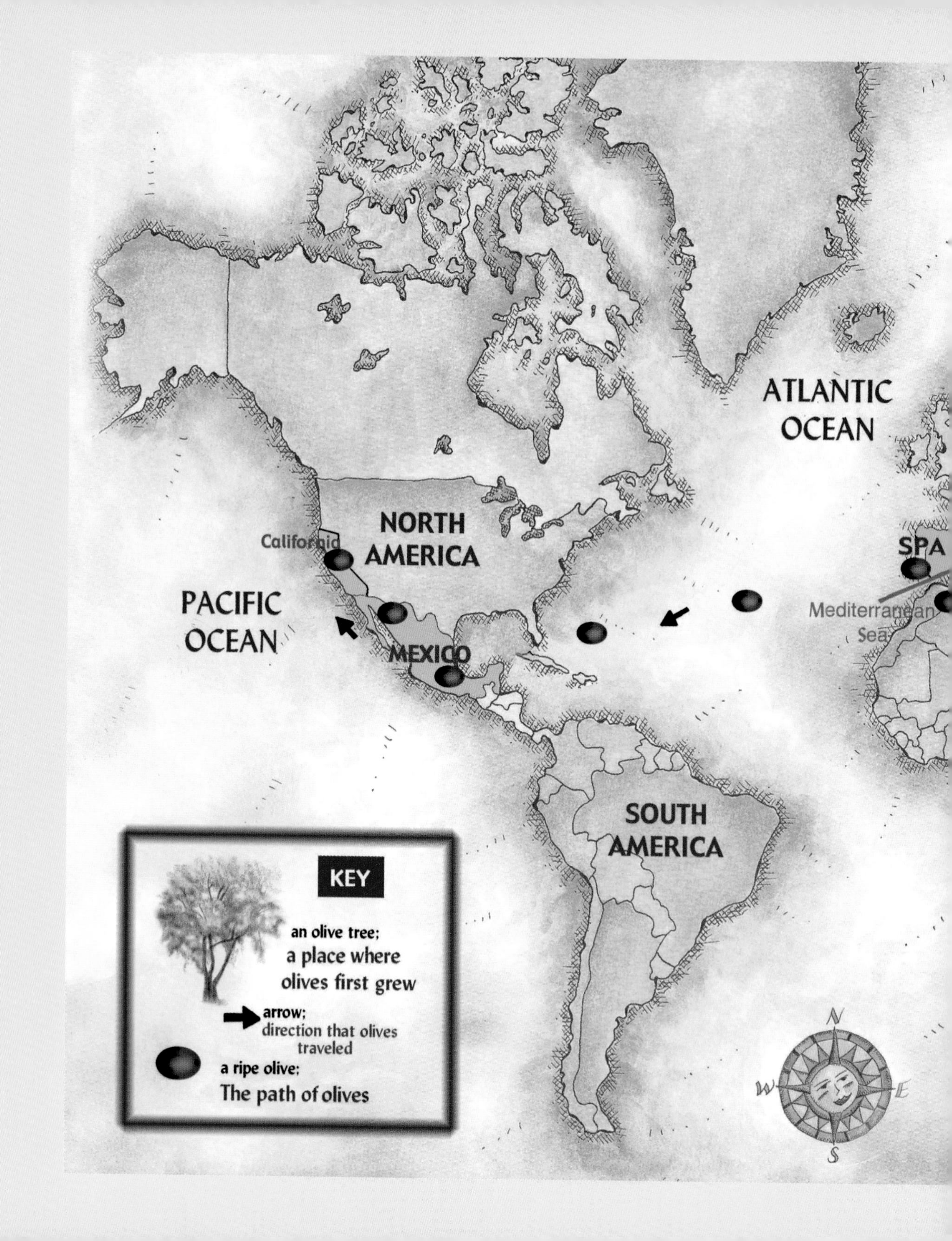

ATLANTIC OCEAN
NORTH AMERICA
California
PACIFIC OCEAN
MEXICO
SPA
Mediterranean Sea
SOUTH AMERICA
KEY
an olive tree;
a place where olives first grew
arrow;
direction that olives traveled
a ripe olive;
The path of olives
N
W
E
S

This map shows where olives came from. Follow the olives to see how they reached the United States.

Grandchild's Gift

Farmers say they plant olive trees for their grandchildren. Why? It can take up to 30 years before the tree grows a good crop, or **harvest**, of olives. When olives first form on the tree, they are nearly white. Young green olives and fully grown black ones are picked for eating and oil.

Farmers could also say that olive trees are gifts for their great-great-great-great grandchildren. There are olive trees that are more than 2,000 years old!

These old olive trees inspired Vincent van Gogh to paint this picture.

a ripe olive

Olive You!

People can grow olives from their seeds, called **pits**. Most people, however, cut a branch from a tree with good olives. Then they plant the branch or attach it to another olive tree.

All olives for eating have to be **cured**. They soak in salt water for a long time before they are eaten. Eating fresh olives is about as tasty as chewing tires, but cured olives are yummy!

Olive tree and olive

Olive of Life

In the old story of Noah's Ark, Noah learns that the flood will end when a dove returns to the ark carrying an olive branch. For thousands of years olives and their oil have meant life to the people of the Mediterranean Sea.

No one knows where and when olives were first planted because it happened so long ago. Nearly 2,500 years ago, the Greeks were writing about the ancient history of olives!

A painting of Noah's Ark

The Perfect Hat

Do you ever wonder what to wear to a fancy party? In ancient Egypt people wore cones that dripped sweet-smelling olive oil on their heads!

Egyptians also ate olives and used the oil for lamps, to keep their skin soft, and as medicine. They were even buried with crowns made of olive leaves. They believed that Isis, the greatest goddess of ancient Egypt, had taught people to grow and use olives.

Egyptian women wearing cones

Greece's Tree of Peace

An old story tells that Poseidon, the god of the sea, and Athena, the goddess of wisdom, both wanted to be the main god of the Greek city of Attica.

Waves crashed when Poseidon gave the people his gift—the first horse. It would help them win wars.

Next Athena planted the first olive tree. It would not help win wars. To the Greeks, the olive tree meant peace. Attica was renamed Athens for the wise goddess.

This vase showing Greek men picking olives is more than 2,000 years old!

Tree of War

Nearly 2,500 years ago an attacking army lit the olive trees of Athens on fire. The old tree that the Greeks believed had been planted by Athena was burned to the ground.

Olive trees are famous for being tough. Soon a green branch grew from the burned roots. When the people of Athens saw it, they knew their city would become great once again.

The Miracle of the Oil

Moses was a great **Jewish** leader who freed his people by leading them out of slavery more than 3,000 years ago. He understood how important olives were to his people. He made laws to protect olive trees.

Olive oil lamps were used to light Jewish temples. The holiday of Hanukkah remembers a special miracle. A lamp with one day's worth of holy olive oil burned for eight days.

A father lights the Hanukkah oil light during the celebration of Hanukkah.

The Blessing of Oil

Olives are also important to **Christians**. Jesus spoke to his people at the Mount of Olives, a hill dotted with olive trees near the city of Jerusalem. He died on a cross made of olive wood.

Even the name Christ has to do with olive oil. It comes from a Greek word. It means someone who has been chosen by God and blessed by holy oil.

Jerusalem, as seen from the Mount of Olives

Holy Light

The **Koran** is the holy book of the **Muslims**. The Muslim religion began in the Middle East, where many olives are grown. In the Koran, the light from olive oil is compared to the holy light of Allah—the Muslim name of God.

Nearly 1,500 years ago, Muslims settled in Spain. Olives were first planted there by the Greeks, but the Muslims planted more. Today more olives are grown in Spain than anywhere else.

A page from a Muslim holy book made in 1558

Queen Olive

An ancient Roman writer called the olive "the queen of all trees." Like Greek winners of the **Olympic** contests, Roman heroes were often crowned with olive branches. Roman workers ate a diet of olives, bread, wine, and salt.

Romans planted olive trees throughout their lands. They believed they were following in the footsteps of Hercules, the strong hero of the old stories. They said he planted olives as he traveled around fighting monsters.

Hercules attacks a monster with help from a friend. Hercules is on the right.

Leonardo's Oil

The first step in making olive oil is to crush the olives. Then they are pressed to squeeze out the oil. Finally, any water or olive pieces are removed from the oil. As the oil is made it is tested. The best tasting oil is called extra virgin olive oil.

There are many ways of making oil, from using stones to machines. Leonardo da Vinci, the famous Italian artist, thought of one. He invented an olive press.

Leonardo da Vinci drew this picture of himself.

Mission Olives

In 1767 Spanish **missionaries**, people who travel to do religious work, brought olive trees from Mexico to California. They came because they wanted the Native Americans to become **Catholic**. They also wanted California to be Spanish.

They built churches and planted olive trees along the coast of California. Some of the trees that they planted still give good harvests today.

This painting shows the San Gabriel Mission of California in 1835.

The Pits!

Olives have long meant peace and joy. Olive oil has been used to crown kings. Romans felt sorry for those who ate bread with butter instead of olive oil.

In Israel, a good person is said to be "pure olive oil." Bad times in America are called "the pits." Good times are the opposite—an olive on each finger and a grin from ear to ear!

Glossary

Catholic (KATH eh lik) — of, or about the Roman Catholic Church (a Christian church)

Christians (KRIS chenz) — believers in the religion based on Jesus Christ

cured (KYOORD) — something kept from going rotten by soaking, salting, or smoking

harvest (HAR vest) — gathering a crop when it is ripe, or a ripe crop of food

Jewish (JOO ish) — of, or about the religion of Judaism

Koran (ke RAN) — the holy book of Islam

missionaries (MISH eh ner eez) — people sent to do religious work

Muslims (MUZ lems) — believers in the religion of Islam

Olympic (oh LIM pik) — of, or about ancient Greek games, or modern sports games

pits (PITS) — the hard seed of some fruits

Index

Further Reading

Brown, Alan and Andrew Langley. *What I Believe.* Millbrook Press, 1999.

Hart, Avery and Paul Mantell *Ancient Greece!: 40 Hands-on Activities to Experience This Wondrous Age.* Williamson Publishing, 1999.

Websites to Visit

www.olympia-oliveoil.com
www.siu.edu/~ebl/leaflets/olive.

About the Author

Melinda Lilly is the author of several children's books. Some of her past jobs have included editing children's books, teaching pre-school, and working as a reporter for *Time* magazine.